Architectural Design Mathematics of Christ Church in Philadelphia, Pennsylvania

by Timothy Scott© 2015

ISBN-13: 978-1512345612

ISBN-10: 151234561X

Table of Contents

1. American Colony Plans

The thirteen colonies sprinkled across the Eastern United States were comprised of people from different origins with accompanying cultures, disparate religious beliefs, and economic pursuits. A summary list follows:

State	Date [Year]	Notes
VA	1607	Business venture named for Elizabeth the virgin Queen
MA	1620	Pilgrims and Puritans
NH	1623	Refugees from MA Puritans
NY	1624	Dutch holdings seized by British and named after Duke of York
NJ	1633	Random settlers
MD	1634	Haven for British Catholics
CONN	1635	Named for *Beside the Long Tidal River*
RI	1636	Roger Williams, founder, Banished from MA for advocating separation of Church and State
DE	1638	Dutch settlers
PA	1643	Dutch and Swede settlers
NC	1653	Loyal nobles to King Charles II were given the land slice off of VA
SC	1670	English settlers from Barbados
GA	1732	Created a buffer to Spain's holdings in the South

Grid designs with order, connectedness, and passage describes the city or colony. The gridded street design has ancient beginnings in Roman and Greek architecture. The design is both replicated and modified as it spread to Europe and to American colonies. Several colony areal views and shapes are displayed on the following pages.

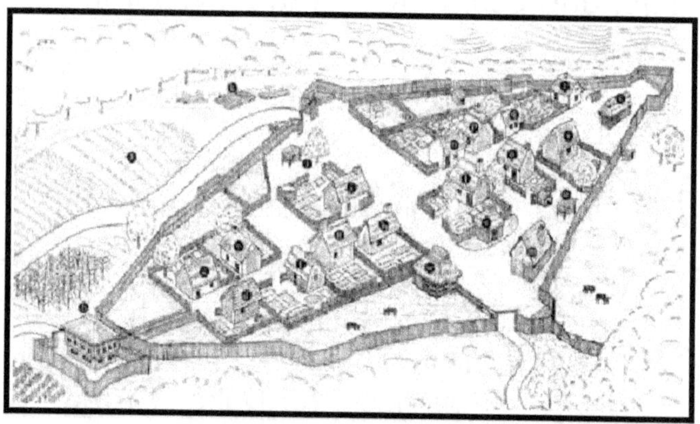

Diamond Plymouth Plantation 1620

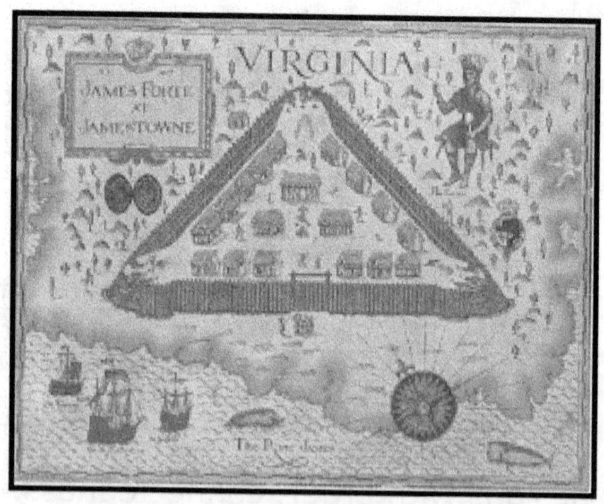

Triangular Jamestown, VA

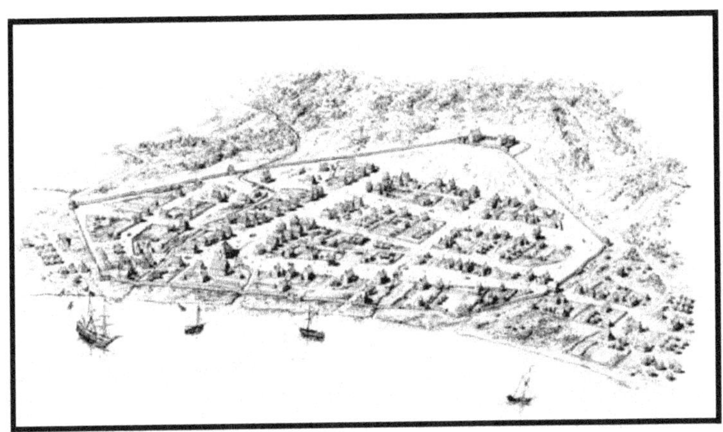

Albany, NY 1686

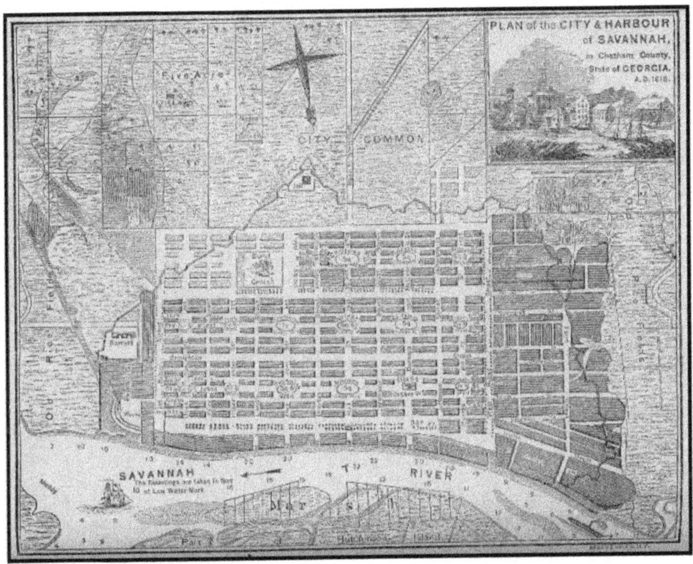

Savannah, GA

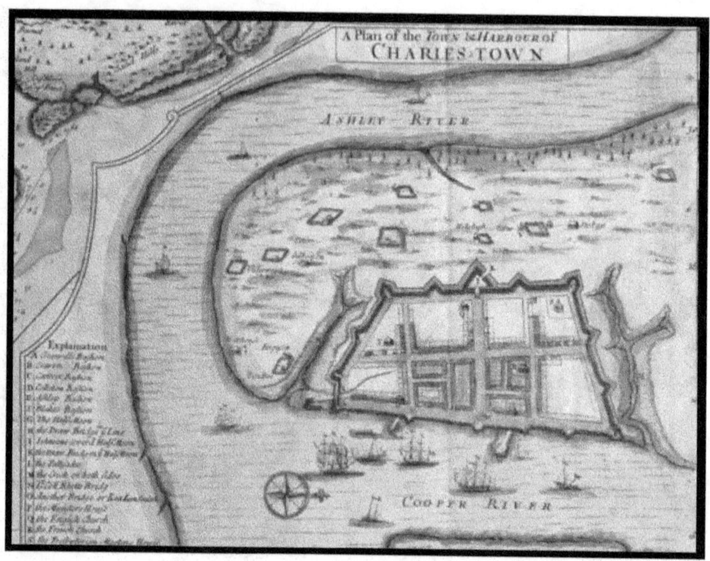

Charlestown, SC

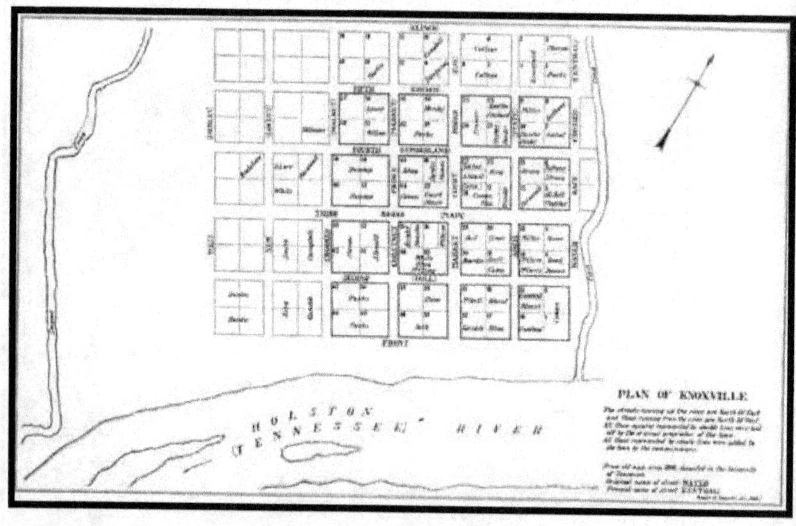

Knoxville, TN 1791

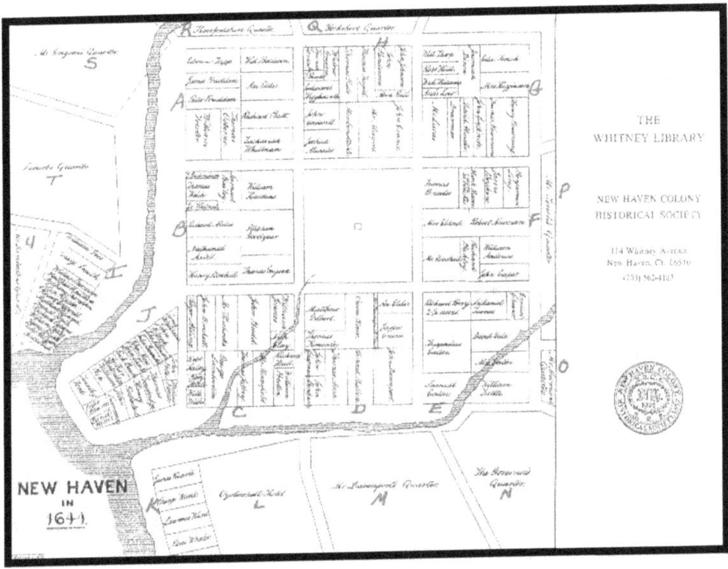

New Haven, CONN Nine Square Arrangement

From the birth of the American colonies religion would step through time in psychological stages of moderation and toleration, to freedom, and finally to freedom of religion. George Washington, who was First President and attended Christ Church of Philadelphia, affirmed that everyman has freedom of conscience to practice their religion as they normally have been. William Penn espoused testimonies of freedom, of education, and of justice; he embraced the idea that education and justice would imply and bring equality. The independence from England, the Constitution, and the first outside celebration of the Eucharist , or mass to re-member Christ, by Catholics took place in Philadelphia.

2. Philadelphia City Plan

Thomas Holme, the General Surveyor of Pennsylvania for William Penn, designed the City of Philadelphia which was to be the capital of Pennsylvania colony in 1863 as reported in the *Portraiture of a City*. *Sylvania* means land of the trees. The rectangular area was subdivided into two squares with both North-South and East-West running streets. The City was planted between two rivers, the Schulkyl and Delaware Rivers which would make it an ideal attractor for European investors for clearly identified plots of land for families, as well as for successful economic futures with the waterway trade route access. The plan envisioned each quadrant of open-space squares of natural park beauty.

The City design has been likened to that of Richard Newcourt who presented a restoration design for London after the great devastation by Black Plague in 1865 and subsequent fire that consumed most of the City the following year. The Philadelphia urban layout presented wider streets for possible evacuation preparedness. Rene Descartes credited with

Cartesian coordinate graphing of ordered pairs of numbers to locate a position in the plane was introduced during 1837. A sketch follows:

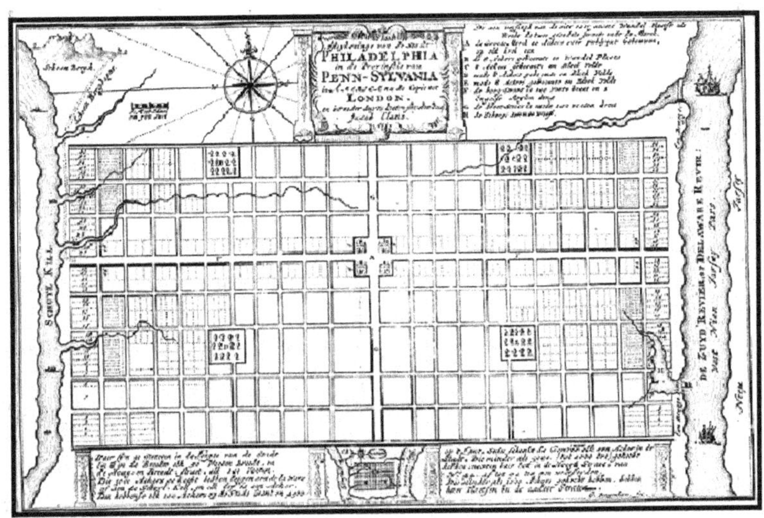

Credit: Philadelphia Meeting Houses

Descriptive phrases for the City of Philadelphia include:

Holy Experiment for William Penn's Religious Quakers

City of Brotherly Love emphasizing religious tolerance

It is a handsome city, but distractingly regular.[Dickens]

On the whole I'd rather be in Philadelphia. [W.C. Fields]

Endeavoring to keep the unity of the Spirit in the bond of peace.

[Ephesians 4: 3]

3. Christ Church Historical Background

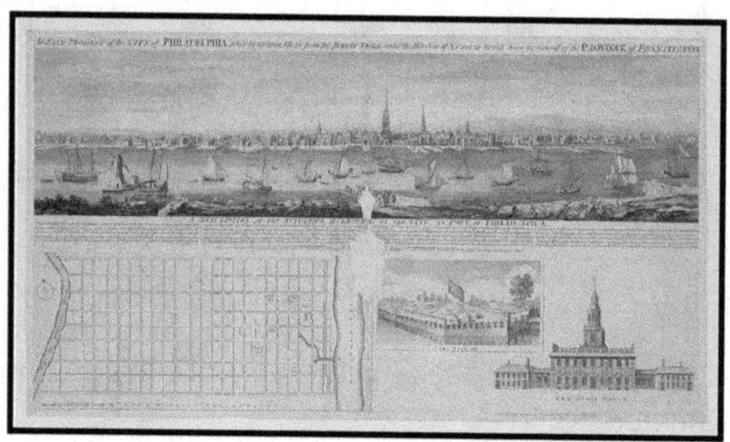

An east prospect of the city of Philadelphia; taken by George Heap from the Jersey shore, under the direction of Nicholas Scull surveyor general of the Province of Pennsylvania / engraved by T. Jefferys.

The Episcopal Church formed its own organization from the Anglican Church separating from the powers of the British King George. Our nation's War of Independence and Constitution is symbolized by the Liberty Bell. The bells at Christ Church of Philadelphia that rang in 1776 are still used today.

The first building was laid in 1695 and aimed for a congregation size of four to five hundred approximately ten percent of the City's population. The architectural history records re-building and enlargements. The present structure dates to approximately 1747 when Dr. Kearsley of the Architecture Committee stated, the *Church is happily finished*. Thomas U. Walter is credited with interior design: Length 90 feet, width 57 feet, and height 47 feet. A chandelier hangs in the center dating from 1744. Walls on the north and west were added in 1756; musical organs or modifications were periodically made [1728, 1763, 1766, 1836, and 1935]; a wine-glass pulpit designed by Folwell in 1770 brought curves to the interior; new pews were added in 1836 [134 on the ground and 36 in the gallery on the second level]. The earliest bell probably hung in the crotch of a tree while later a ring of eight bells were ordered from London in 1754 weighing about nine thousand pounds in total. The largest bell weighs two thousand and forty pounds and the smallest five hundred pounds. The steeple designed by Robert Smith above the tower building raised it to a height of one hundred and ninety-six feet. This was the tallest structure for nearly a century. The baptismal font used for William Penn in 1644 at All Hallows Church in London was sent to the Church in 1697.

"We utterly deny all outward wars and strife and fightings with outward weapons, for any end or under any pretence whatsoever. And this is our testimony to the whole world. The spirit of Christ, by which we are guided, is not changeable, so as once to command us from a thing as evil and again to move unto it; and we do certainly know, and so testify to the world, that the spirit of Christ, which leads us into all Truth, will never move us to fight any war against any man with outward weapons, neither for the kingdom of Christ, nor for the kingdoms of this world.

[An early version of Penn's Quakers or Religious Society of Friends *Peace Testimony* was in 1661, in a declaration to King Charles II of England.]

4. Architecture Design Principles and Mathematics

4a. Arithmetic, Geometry, and Musical Scales

External and internal structural revelations with spire, light transmissions and reflections

"The whole appearance is chaste, simple, and elegant; and there is throughout a quiet grandeur and sublimily, which at once inspire sentiments of devotion and awe." [Benjamin Dorr, *Historical Account of Christ Church From Its Foundation (1859)]*

Numbers and ratios in the geometry of the Church are likened to the ABC's of a language[1]. The numbers, of course, express specific values; however, there are meta-number ideas with associations.

One: Unity; Two: Duality, symmetry, or co-existence of parts; Three: Trinity, triangle, as a roof shape. The numbers two and three are prime numbers which are like the building blocks of all integers. Karl Gauss proved this Fundamental Theorem of Arithmetic for representing positive integers as multiplication factors of primes.

The numbers one, two, three both add and multiply yielding the same number six.

$$1 + 2 + 3 = 6$$

and $1 \times 2 \times 3 = 6$

The six sided polygon, or hexagon, has been likened to the geometry figures of ancients on the flower or seed of life drawn by compass.

The number seven may be viewed as demarking a balance between the product of numbers one through six and eight through ten.

$$1 \times 2 \times 3 \times 4 \times 5 \times 6 =$$

$8 \times 9 \times 10$ that is,

$$720 = 720.$$

This mathematical equation is a statement of equality or restoring balance between the two sides. The word origin for algebra derives from ideas of balance, equality, and restoration.

Mathematics of the Golden proportionexpresses equality and balance, as well as that of the musical octave, or halving string-length and doubling frequency, and fifths (ratios of three to two and eight to five and its reciprocal) are displayed in the church's design. For example, the original length, 90, width, 57,and height , 47, in units of feet, are numbers to obtain:

$$\frac{47}{90} = 0.52 \approx \frac{1}{2},$$

$$\frac{90}{57} = 1.58 \approx 1.6$$

Further quantitative musical ratios and Fibronacci series, such as 1, 1, 2, 3, 5, 8, are discussed and listed in the appendices.

Clear glass Palladian windows in Golden portions enlighten the altar and Church. The Palladian window design consists of a three part window attributed to architect Andrea Palladio. The smaller windows have $27 = 3^3 = 3 \cdot 3 \cdot 3$ glass panes and the total number of panes including these of the larger central window is 176, which is coincidently associated with the year 1776 of the War for

His radiance is like sunlight; He has rays flashing from His hand,
And there is hidings of His power. [Habakuk 3:4]

Independence of the Nation for which prominent members of the church were actively involved.

Beauty will result from the form and correspondence of the whole, with
Respectto the several parts with regard for each other, and these again to
the whole; that the structure may appear an entire and complete body, wherein
each member agrees with the other, and all necessary to compare what you
intend to form.

From Andrea Palladio in *Quattro Libri dell' Architettura*[Four Books of Architecture] first published in Venice in 1570.

4b. Golden Balance

The rectangular floor where communion service is conducted has approximate Golden proportion ratio dimensions.

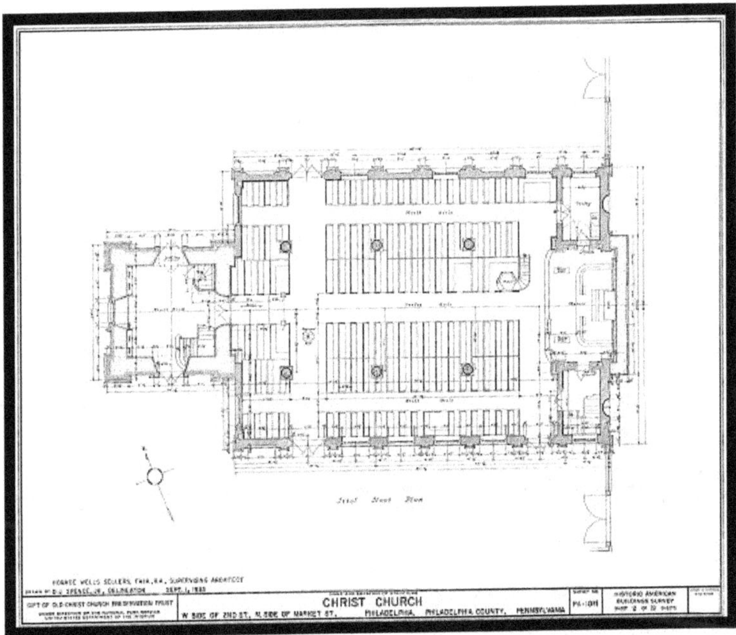

1. Gematria is the actual association of a number with an alphabet letter. For example, a most holy name for God, YHWH, translates to the number 26

and the Creator, Alpha and Omega translates to 801.

The idea of total balance accompanying its construction or formulation of the Golden proportion is noted to emphasize that it is not just a numerical value.

The exponentials of *phi* in the window dimensions as been associated with the lengths of the digits of a human's hand.

Father, into your hands I commit my Spirit
[Luke 23:46]

The name Israel means people struggling with GOD. The Jews were to be the peacemakers and priests of the Ten Commandments [Benjamin Efron, *The Message of the Torah*].

5. Concluding Remarks

Assembling people who give the same answer to the question posed by Jesus of Nazareth, *Whom say you that I am ?[Mathew 14:15-18]* as did Simon Barjona, [...*I say unto you, that you are Peter, and upon this rock I will build my church; and the gates of hades shall not prevail against it.*] has been accomplished at churches like Christ Church of Philadelphia and of St. Peter's. The results of this investigation on a building's design messages by vision and form, as speech conveys connotative meanings and aesthetics to words or musical consonance derive from music, or Fibronacci numbers, and a Golden proportion.

The religious concepts of oneness, or unison, as a unification of a church koininia, or community, is embodied in the large worship space. The dualities of an individual's blessings and sorrows are brought together through Biblical readings, prayer, and singing of hymns. Furthermore, harmony by the musical scale ratios of octaves, fifths, and thirds are noted; in addition to, the trinity of Palladium windows' exterior. The mathematical design aspects of arithmetic are conducive to living and worshipping in simplicity.

The Golden ratio appears in the floor dimensions composing a perfect balanced design area of an octave to celebrate life in the light, in peace within the body, passing through the Palladium windows. The window panes give an easy association and memory to the year 1776 and war for independence. The window dimensions appear to approximate powers of this balance ratio, called *phi* which are likened to a hand's digits.

Mathematics' aims for true statements and truth coincides with religion. The philosophies of truth may be characterized by consistency theory, coherence, pragmatism, existentialism, and relativity. Bible scholars present these theories on truth from Christian scripture.

6. References and Information Web-links

1). Nash, Gary B., *The American People: Creating A Nation* (7th Edition), Pearson (2010).

2). Peterson, Charles E., *Building Early America: Contributions Toward the History of a Great Industry*, Astragal Pr. (1992)

3). Ifrah, Georges, *The Universal History of Numbers*, Wiley (1998).

4). Measured Drawing onwww.loc.gov

5). The Building of Christ Church online at www.christchurchphila.org

6). Efron, Benjamin, *The Message of the Torah*, Ktav Pub. (1963).

7). Huntley, H.E., *The Divine Proportion*, Dover (1970).

8). McDowell, Josh, *The New Evidence That Demands A Verdict: I & II Fully Updated in One Volume To Challenge Questions Challenging Christians in the 21st century*, Thomas Nelson (1999).

9). Livio, Mario, 2003 *The Golden Ratio: The Story of Phi, The World's Most Astonishing Number,* Broadway.

10). Binsfield, John, The Organ and Bells of Christ Church Philadelphia, Christ Church Preservation Trust.

7. Appendices

7a. Appendix A: Golden Balance Construction By Compass And Ruler:

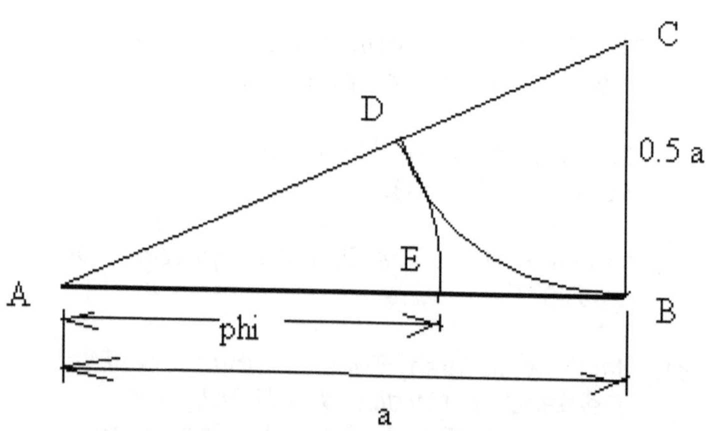

In Euclid's *Elements* consisting of thirteen books on geometry, a problem is given on dividing a line into its mean and extreme ratio [Book 6 Proposition 30]. The letter phi was chosen to credit Phidias, the Greek sculptor of statues for the Parthenon.

Simple Derivation Of The Golden Proportion, denoted by "Phi" or in Greek, the letter φ.

Here, proportion is articulated as the equality of two ratios in the form

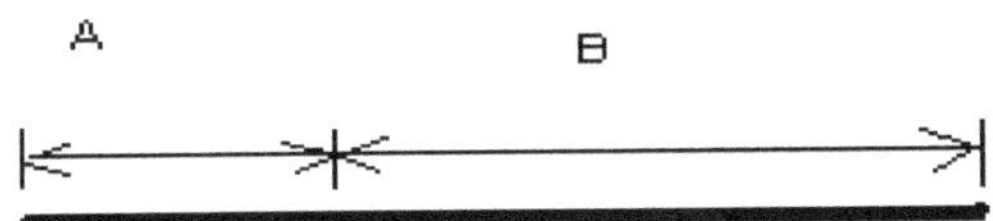

of line segment lengths. $\frac{A}{B} = \frac{B}{A+B}$

Using algebra and the properties of Real Numbers one may define $x = \frac{B}{A}$, and obtain the following equation to solve,

$$x^2 - x - 1 = 0$$

The two solutions are,

$$x = \frac{1+\sqrt{5}}{2} \approx 1.61803$$

Or $$x = \frac{1-\sqrt{5}}{2}$$

which are intercepts of the parabola equation.

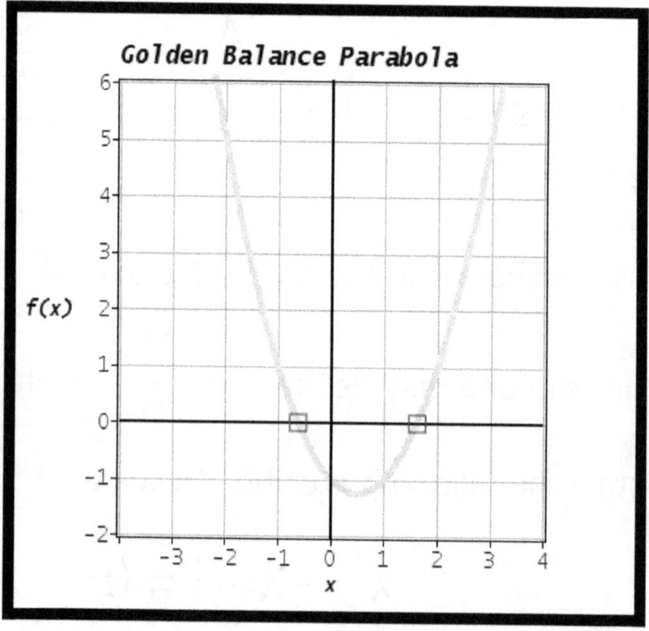

The Greek letter *phi*, ϕ, is often used to represent the number named after the sculptor Phidias for his fifth century classical Greek works in the Parthenon.

The Golden balance may be represented as a continuing fraction,

$$\emptyset = 1 + \cfrac{1}{1 + \cfrac{1}{1+\cfrac{1}{1+\cfrac{1}{1+\cdots}}}}$$

|

Additionally, the ratio appears as a limit of the Fibronacci series which starts with the numbers one and two, and continues by addition of the previous two numbers in the calculated series. Fibronacci, abbreviation of Filius Bonaccio, in 1202 wrote Liber Abaci as an arithmetic and algebra treatise. The Fibronacci series is formed from the integers by the rule: let the first two terms be given by 1, 1. Then the next terms are given recursively by summing the previous two terms. The number series becomes 1, 1, 1+1=2, 1+2=3, 2+3=5, 3+5=8, 5+8=13,....

The color yellow in the human eye's response is approximately this ratio or proportion; hence referred to here as *Golden Portion* of the light. Face and human body measurements display Golden ratios of lengths.

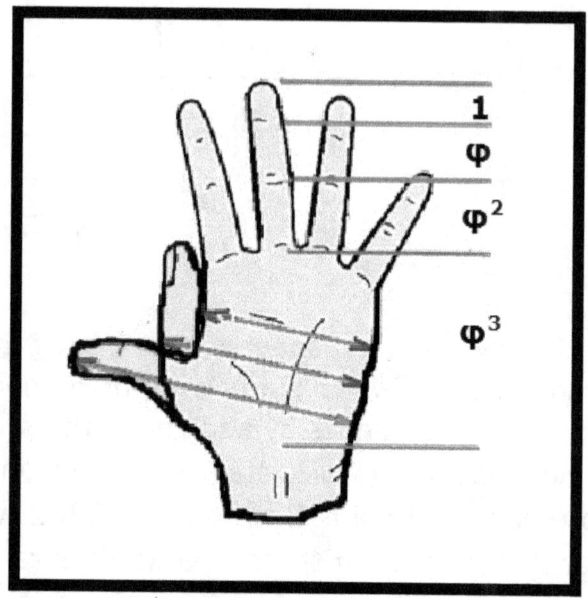

Digits of a hand

7b. Appendix B: George Washington's Letter

For most of the Colonial era, *tolerance* meant tolerance for different denominations of Christians, and sometimes Catholics. In this famous letter, Washington declared that American tolerance included Jews, too. Mikva Israel, the oldest synagogue in Philadelphia, annually have a dinner together with Christ Church's members.

Letter to the Congregation in Newport, Rhode Island, *January 1, 1790*

While I received with much satisfaction your address replete with expressions of esteem, I rejoice in the opportunity of assuring you that I shall always retain grateful remembrance of the cordial welcome I experienced on my visit to Newport from all classes of citizens.

The reflection on the days of difficulty and danger which are past is rendered the more sweet from a consciousness that they are succeeded by days of uncommon prosperity and security.

If we have wisdom to make the best use of the advantages with which we are now favored, we cannot fail, under the just administration of a good government, to become a great and happy people.

The citizens of the United States of America have a right to applaud themselves for having given to mankind examples of an enlarged and liberal policy—a policy worthy of imitation. All possess alike liberty of conscience and immunities of citizenship.

It is now no more that toleration is spoken of as if it were the indulgence of one class of people that another enjoyed the exercise of their inherent natural rights, for, happily, the

Government of the United States, which gives to bigotry no factions, to persecution no assistance, requires only that they who live under its protection should demean themselves as good citizens in giving it on all occasions their effectual support.

It would be inconsistent with the frankness of my character not to avow that I am pleased with your favorable opinion of my administration and fervent wishes for my felicity.

May the children of the stock of Abraham who dwell in this land continue to merit and enjoy the good will of the other inhabitants—while everyone shall sit in safety under his own vine and fig tree and there shall be none to make him afraid.

May the father of all mercies scatter light, and not darkness, upon our paths, and make us all in our several vocations useful here, and in His own due time and way everlastingly happy.

G. Washington

7c. Appendix C: Quadratic Formula Derivation

Solve, $ax^2 + bx + c = 0$

$$\frac{ax^2 + bx + c}{a} = \frac{0}{a}$$

$$\frac{ax^2}{a} + \frac{b}{a}x + \frac{c}{a} = 0$$

$$x^2 + \frac{b}{a}x + \frac{c}{a} = 0$$

$$x^2 + \frac{b}{a}x + \left(\frac{1}{2}\frac{b}{a}\right)^2 = \left(\frac{1}{2}\frac{b}{a}\right)^2 - \frac{c}{a}$$

$$\left(x + \frac{1}{2}\frac{b}{a}\right)^2 = \left(\frac{1}{2}\frac{b}{a}\right)^2 - \frac{c}{a}$$

$$\left(x + \frac{1}{2}\frac{b}{a}\right)^2 = \frac{b^2}{4a^2} - \frac{4ac}{4a^2}$$

$$\left(x + \frac{1}{2}\frac{b}{a}\right) = \pm\sqrt{\frac{b^2 - 4ac}{4a^2}}$$

$$\left(x + \frac{1}{2}\frac{b}{a}\right) = \pm\frac{\sqrt{b^2 - 4ac}}{\sqrt{4a^2}}$$

$$\left(x + \frac{1}{2}\frac{b}{a}\right) = \pm\frac{\sqrt{b^2 - 4ac}}{\sqrt{4}\sqrt{a^2}}$$

$$\left(x + \frac{1}{2}\frac{b}{a}\right) = \pm\frac{\sqrt{b^2 - 4ac}}{2a}$$

$$x = -\frac{1}{2}\frac{b}{a} \pm \frac{\sqrt{b^2 - 4ac}}{2a}$$

or finally the quadratic formula,

$$\boldsymbol{x = \frac{-b \pm \sqrt{b^2 - 4ac}}{2a}}$$ solves

$$\boldsymbol{ax^2 + bx + c = 0}$$

7d. Appendix D: Musical Ratios

The equal temperate scale of twelve notes equally spaced on an octave in exponent notation is shown in the table. Notations of fractions of positive integers, prime number

	Ratio			
C	1/1	1/1	$\sqrt[12]{2^0}$	$(2^0)^{\frac{1}{12}}$
C#	$\frac{15}{16}$	$\frac{3\cdot5}{2^4}$	$\sqrt[12]{2^1}$	$(2^1)^{\frac{1}{12}}$
D	8/9	$\frac{2^3}{3^2}$	$\sqrt[12]{2^2}$	$(2^2)^{\frac{1}{12}}$
D#	5/6	$\frac{5}{2\cdot3}$	$\sqrt[12]{2^3}$	$(2^3)^{\frac{1}{12}}$
E	4/5	$\frac{2^2}{5}$	$\sqrt[12]{2^4}$	$(2^4)^{\frac{1}{12}}$
F	3/4	$\frac{3}{2^2}$	$\sqrt[12]{2^5}$	$(2^5)^{\frac{1}{12}}$
F#	5/7	$\frac{5}{7}$	$\sqrt[12]{2^6}$	$(2^6)^{\frac{1}{12}}$
G	2/3	$\frac{2}{3}$	$\sqrt[12]{2^7}$	$(2^7)^{\frac{1}{12}}$
G#	5/8	$\frac{5}{2^3}$	$\sqrt[12]{2^8}$	$(2^8)^{\frac{1}{12}}$
A	3/5	$\frac{3}{5}$	$\sqrt[12]{2^9}$	$(2^9)^{\frac{1}{12}}$
A#	4/7	$\frac{2^2}{7}$	$\sqrt[12]{2^{10}}$	$(2^{10})^{\frac{1}{12}}$

B	8/15	$\frac{2^3}{3\cdot5}$	$\sqrt[12]{2^{11}}$	$(2^{11})^{\frac{1}{12}}$
C	1/2	$\frac{1}{2}$	$\sqrt[12]{2^{12}}$	$(2^{12})^{\frac{1}{12}}$

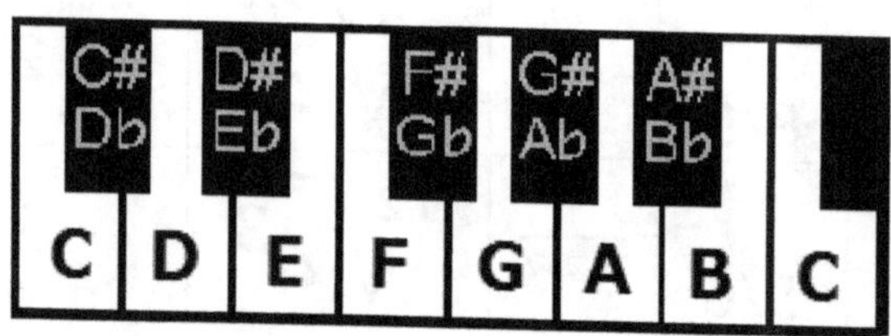

factorizations, or bases and exponents are listed. A logarithm representation which is an inverse function to the exponential function may be used as well. Ernst Weber was first to use logarithms for quantitative human response to physical stimulus followed by Gustav Weber's alternative representation of stimulus perception relationships. Differences in scales include a pentatonic scale of five notes, a whole tone scale of six notes, a diatonic seven note scale, as well as, octatonic eight and other ancient scales of sixteen and twenty four notes. Calculated frequencies are calculated with these multipliers when specific notes are fixed to a frequency such as 440 Hertz for A.

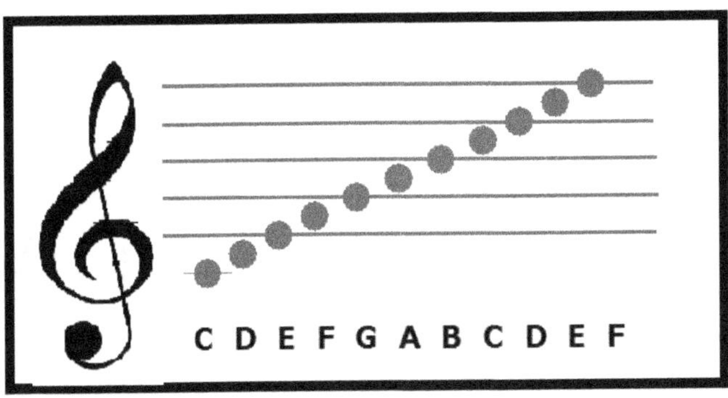

7e. Appendix E. Rules of Bases and Exponents

$$x^n = x \cdot x \cdot x \ldots\ldots x$$

n multiplication factors
(exponent or power n) of the base x

$$x^0 = 1$$

$$x^1 = x$$

$$x^m \cdot x^n = x^{m+n}$$ Product Rule

$$\left(\frac{x^m}{x^n}\right) = x^{m-n}$$

Quotient Rule

$$\left(x^m\right)^n = x^{m\cdot n} = x^{n\cdot m} = \left(x^n\right)^m$$

Power Rule

$$x^{-n} = \frac{1}{x^n}$$

[Negative exponents for Fractions]

$$\sqrt[n]{x} = x^{\frac{1}{n}}$$

[Exponent equivalent of Radical Notation]

$$\left(x \cdot y\right)^n = x^n y^n$$

The logarithm definition is shown as the inverse function $f(x)$ to the exponential,

$$f(x) = b^x,$$

$$log_b(b^x) = x$$

7f. Appendix F. Sound Intensity, Decibels, and Logarithms

The intensity of the sound given on a measured weighted decibel, dB, scales given by the formula,

$$dB = 10 \cdot log\left(\frac{p^2}{p_0^2}\right).$$

Rewriting using a property of logarithms this may be written,

$$dB = 20 \cdot log\left(\frac{p}{p_0}\right)$$

where $p,\ and p_0$ are the measured and reference pressures, respectively.

The loudness of multiple sound sources near each other , thereby considered as a point source, is given by the summation of their pressures rather than addition of the decibel values. That is, pressure is given in terms of decibels by,

$$p_0 \cdot 10^{\frac{dB}{20}} = p$$

In air, use of a sound wave with a pressure of 20 micro-Pascal as the reference is made. This value for sound pressure in air of 20 μPa corresponds to sounds at a frequency of 1000 Hz that can just be heard by humans.

Properties of Logarithms (with base b) are listed below. Natural logarithms are written to the base e, written as ln for $\log_e$

$$\log_b(c \cdot d) = \log_b c + \log_b d$$

$$\log_b\left(\frac{c}{d}\right) = \log_b c - \log_b d$$

$$\log_b(c)^n = n \cdot \log_b c$$

$$\log_a x = \frac{\log_b x}{\log_b a}$$

Exponential x^n and Logarithmic Functions

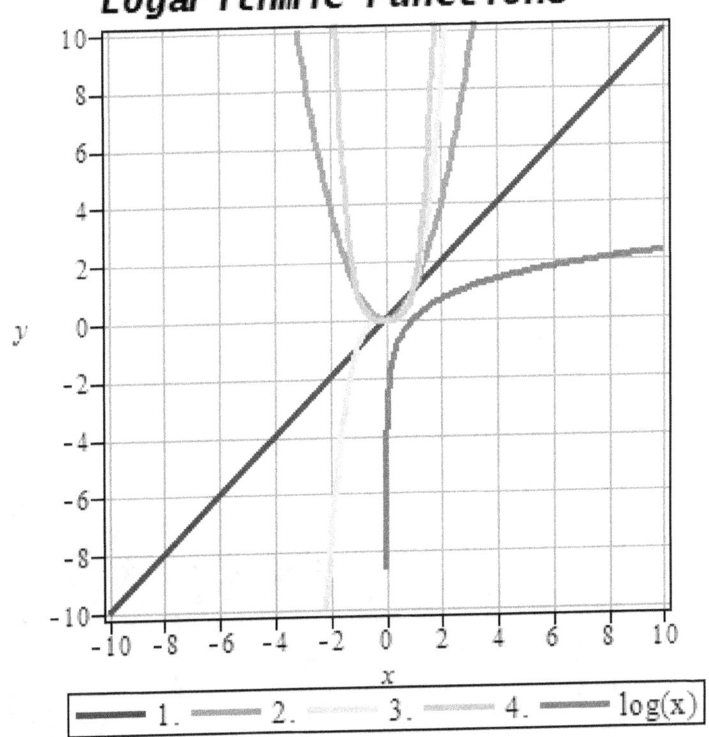

7h. Appendix H: Selected Prints and Drawings of Christ Church

Detroit Photographic Co.

The East Prospect of the City of Philadelphia in the Province of Pennsylvania drawn for the London Magazine.

Designed and constructed by John Folwell.

Set of 8 bells cast in England in 1754 at the White chapel Foundry.

Magna Carta window depicting First Prayer of Congress with Jacob Duché. The Liberty Window, reproduces Matteson's painting: "Prayer in the First Congress, Sept. 1774."

Christ Commissioning the Apostles

Photograph of Early Christ Church Interior

Painting by W. Mason 1837 of the interior of Christ Church with the organ in the gallery.

Selected Architecture Buildings Survey Drawings at the Library of Congress:

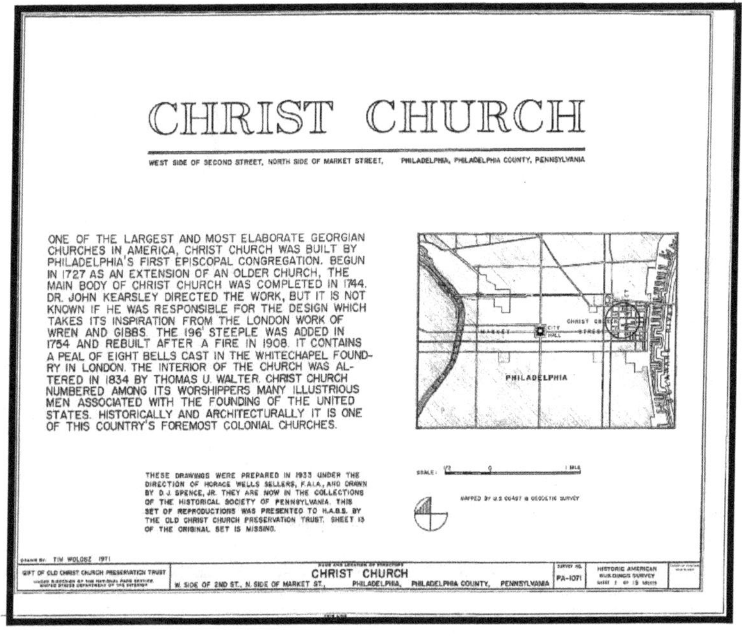

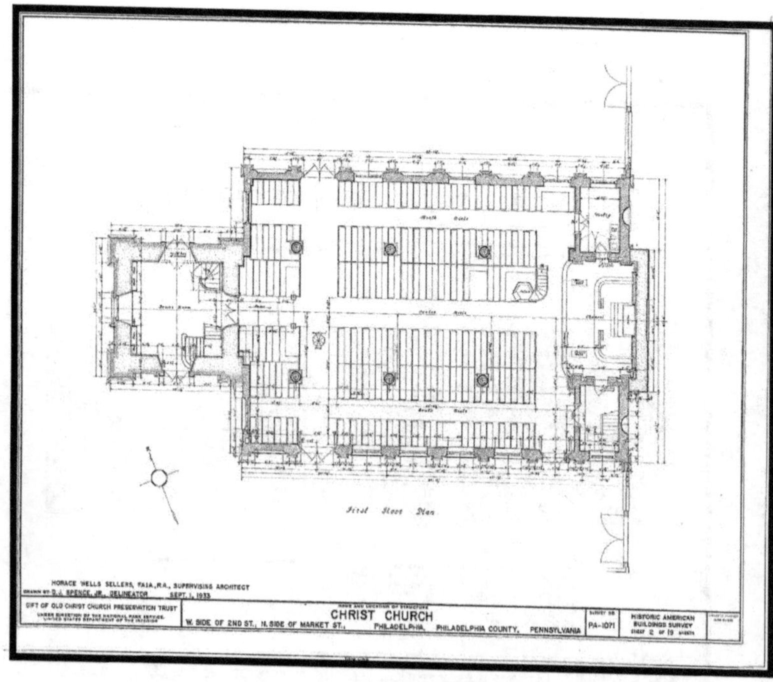
First Floor Plan
HORACE WELLS SELLERS, FAIA, R.A., SUPERVISING ARCHITECT
D. J. SPENCE, JR., DELINEATOR SEPT. 1, 1933
GIFT OF OLD CHRIST CHURCH PRESERVATION TRUST
CHRIST CHURCH
W. SIDE OF 2ND ST., N. SIDE OF MARKET ST., PHILADELPHIA, PHILADELPHIA COUNTY, PENNSYLVANIA
PA-1071
HISTORIC AMERICAN BUILDINGS SURVEY
SHEET 2 OF 19 SHEETS

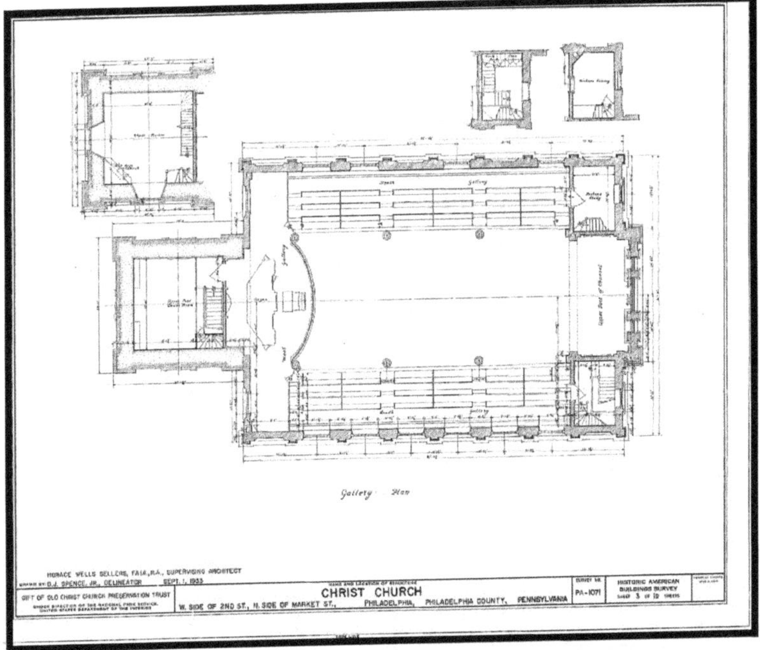
Gallery Plan
HORACE WELLS SELLERS, FAIA., R.A., SUPERVISING ARCHITECT
D.J. SPENCE, JR., DELINEATOR
SEPT. 1, 1933
GIFT OF OLD CHRIST CHURCH PRESERVATION TRUST
CHRIST CHURCH
W. SIDE OF 2ND ST., N. SIDE OF MARKET ST., PHILADELPHIA, PHILADELPHIA COUNTY, PENNSYLVANIA
PA-1071
HISTORIC AMERICAN BUILDINGS SURVEY
3
19

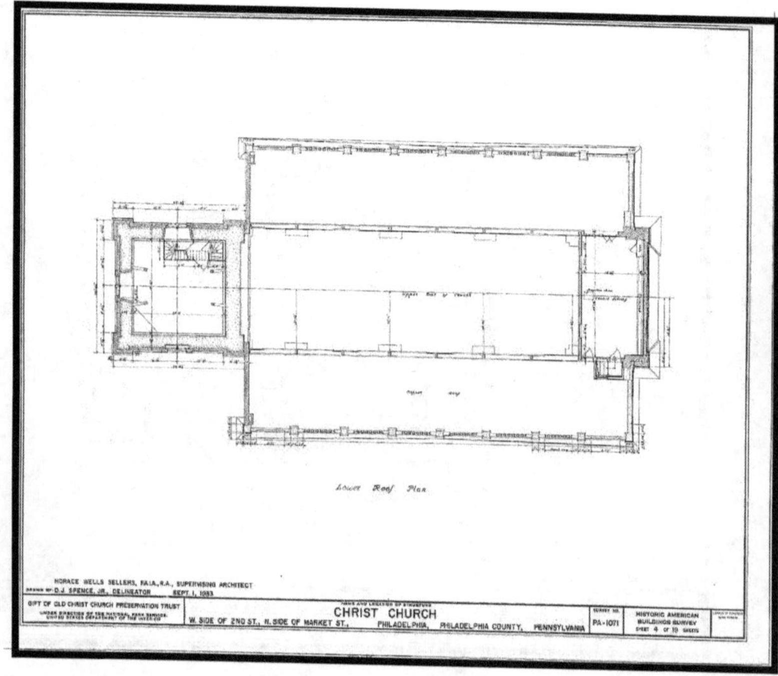
Lower Roof Plan
HORACE WELLS SELLERS, F.A.I.A., R.A., SUPERVISING ARCHITECT
D.J. SPENCE, JR., DELINEATOR
SEPT. 1, 1983
GIFT OF OLD CHRIST CHURCH PRESERVATION TRUST
CHRIST CHURCH
W. SIDE OF 2ND ST., N. SIDE OF MARKET ST., PHILADELPHIA, PHILADELPHIA COUNTY, PENNSYLVANIA
PA-1071
HISTORIC AMERICAN BUILDINGS SURVEY
SHEET 4 OF 19 SHEETS

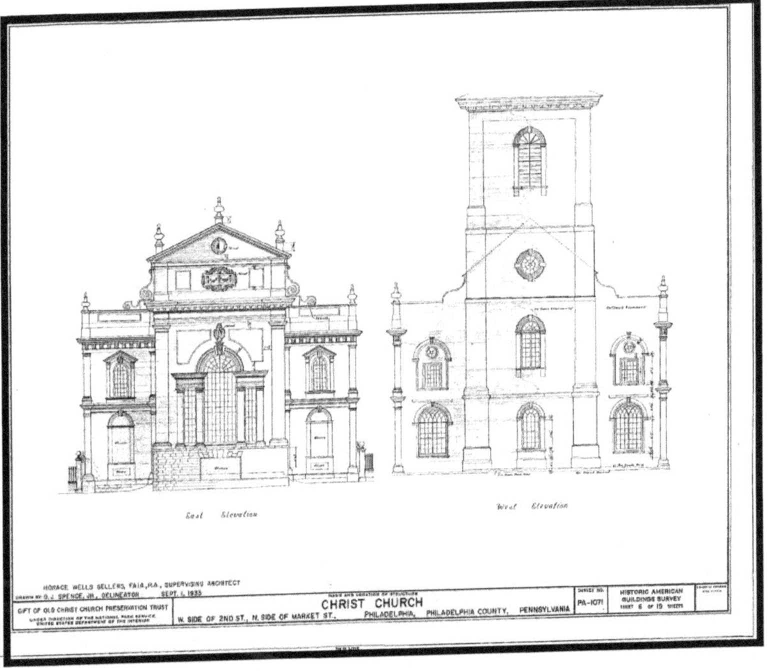
East Elevation
West Elevation
HORACE WELLS SELLERS, FAIA, R.A., SUPERVISING ARCHITECT
DRAWN BY O. J. SPENCE, JR., DELINEATOR SEPT. 1, 1935
GIFT OF OLD CHRIST CHURCH PRESERVATION TRUST
UNDER DIRECTION OF THE NATIONAL PARK SERVICE
UNITED STATES DEPARTMENT OF THE INTERIOR
NAME AND LOCATION OF STRUCTURE
CHRIST CHURCH
W. SIDE OF 2ND ST., N. SIDE OF MARKET ST., PHILADELPHIA, PHILADELPHIA COUNTY, PENNSYLVANIA
SURVEY NO.
PA-1071
HISTORIC AMERICAN BUILDINGS SURVEY
SHEET 6 OF 19 SHEETS

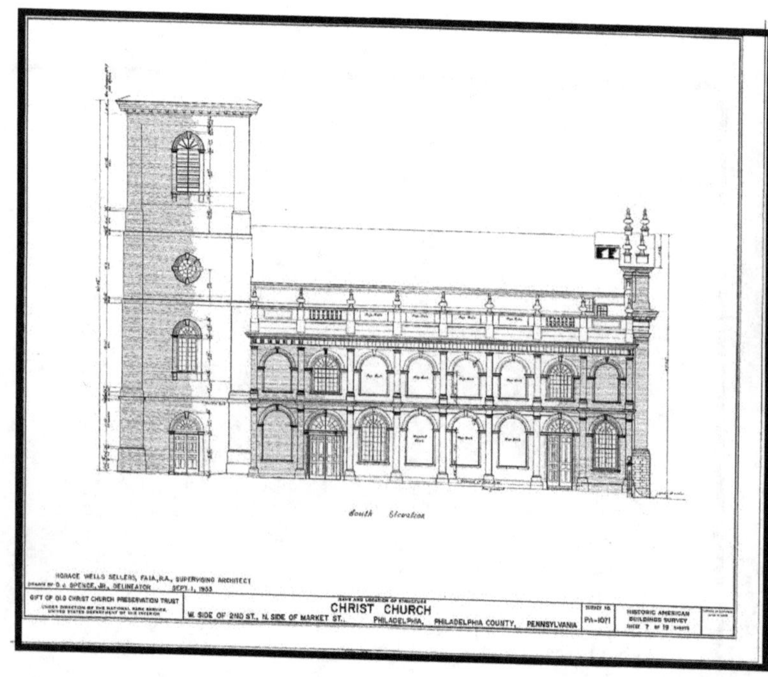
South Elevation
HORACE WELLS SELLERS, FAIA, R.A., SUPERVISING ARCHITECT
D. J. SPENCE, JR., DELINEATOR
SEPT. 1, 1953
GIFT OF OLD CHRIST CHURCH PRESERVATION TRUST
CHRIST CHURCH
W. SIDE OF 2ND ST., N. SIDE OF MARKET ST.
PHILADELPHIA, PHILADELPHIA COUNTY, PENNSYLVANIA
PA-1071
HISTORIC AMERICAN BUILDINGS SURVEY
SHEET 7 OF 19 SHEETS

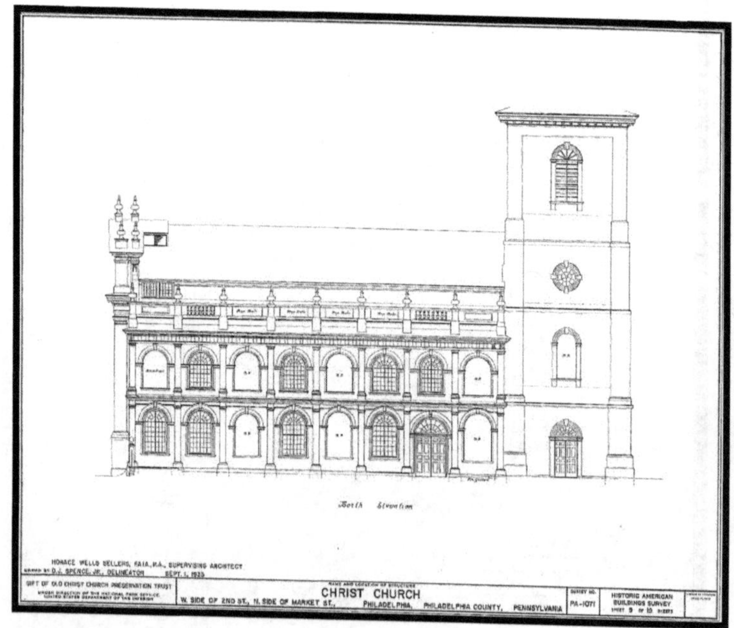
North Elevation
HORACE WELLS SELLERS, FAIA, MA., SUPERVISING ARCHITECT
D.J. SPENCE, JR., DELINEATOR
SEPT. 1, 1935
GIFT OF OLD CHRIST CHURCH PRESERVATION TRUST
CHRIST CHURCH
W. SIDE OF 2ND ST., N. SIDE OF MARKET ST.,
PHILADELPHIA, PHILADELPHIA COUNTY, PENNSYLVANIA
PA-1071
HISTORIC AMERICAN
BUILDINGS SURVEY
SHEET 9 OF 10 SHEETS

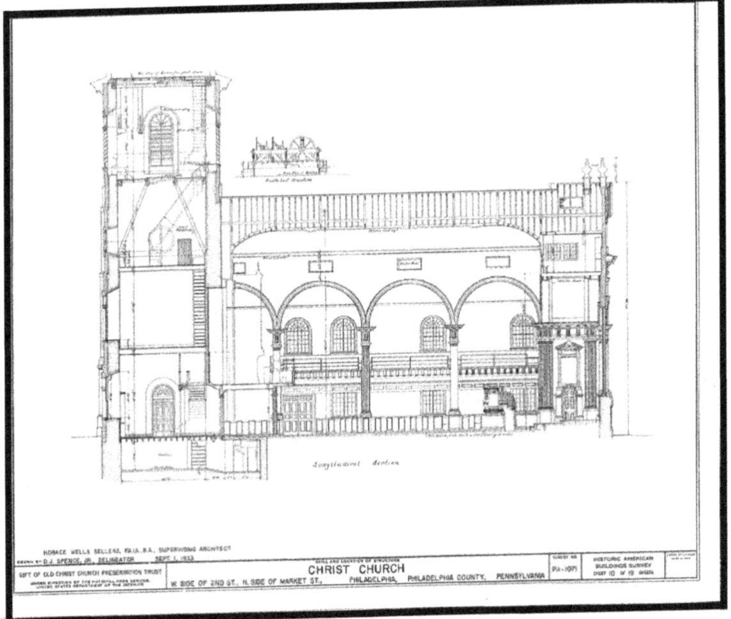
Longitudinal Section
HORACE WELLS SELLERS, F.A.I.A., B.A., SUPERVISING ARCHITECT
D.J. SPENCE, JR., DELINEATOR
GIFT OF OLD CHRIST CHURCH PRESERVATION TRUST
CHRIST CHURCH
W. SIDE OF 2ND ST., N. SIDE OF MARKET ST., PHILADELPHIA, PHILADELPHIA COUNTY, PENNSYLVANIA
HISTORIC AMERICAN BUILDINGS SURVEY

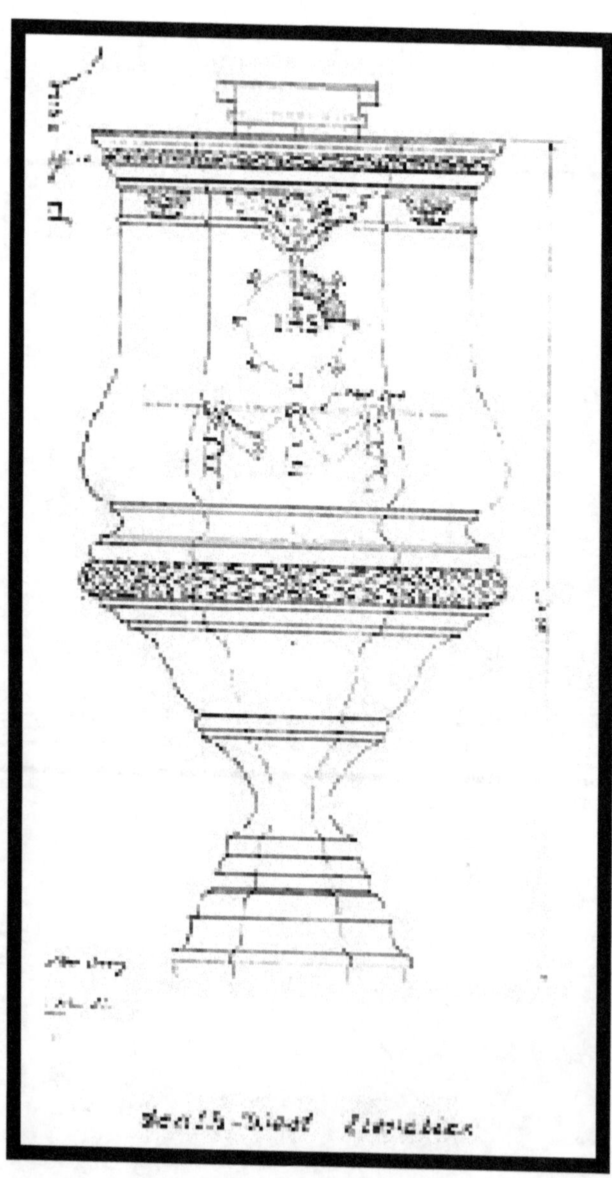

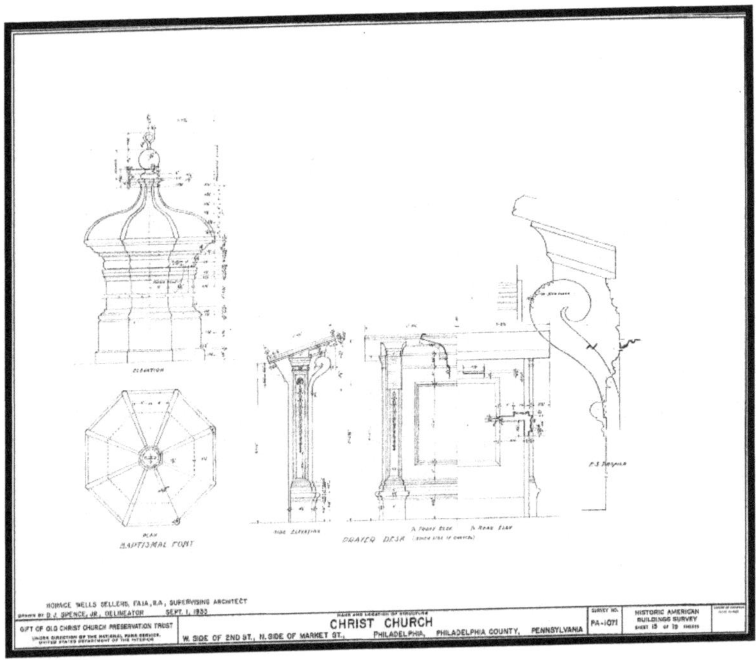

Contributor: Historic American Buildings Survey - Walter, Thomas U - Hamilton, Andrew - Hewitt, G W - Price, Virginia B - Macioge, E - Massey, J - Biand, L - Cooper, R

H. Flower of Life Geometry Drawing

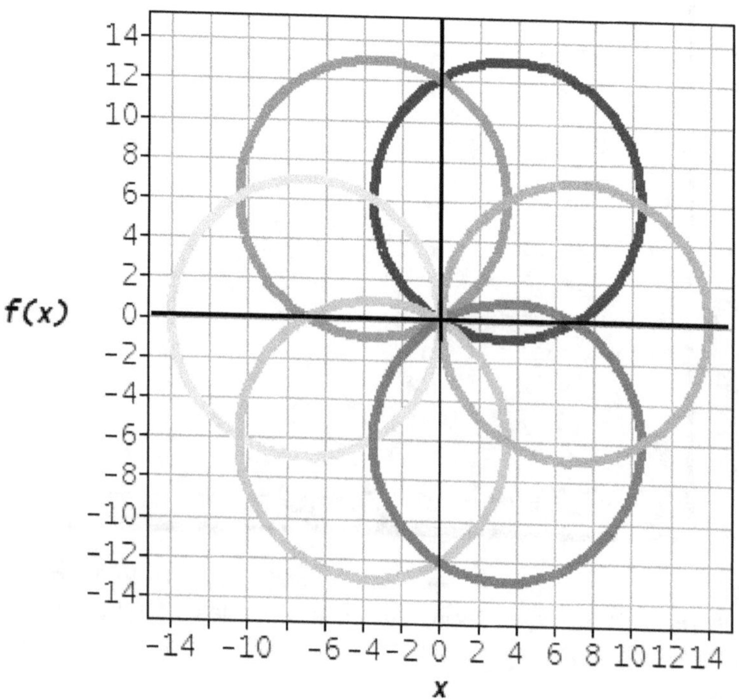

Equation for a circle of radius r and centered at the point (h, k) in the plane is given by, $(x - h)^2 + (y - k)^2 = r^2$

Flower with Seven Petals

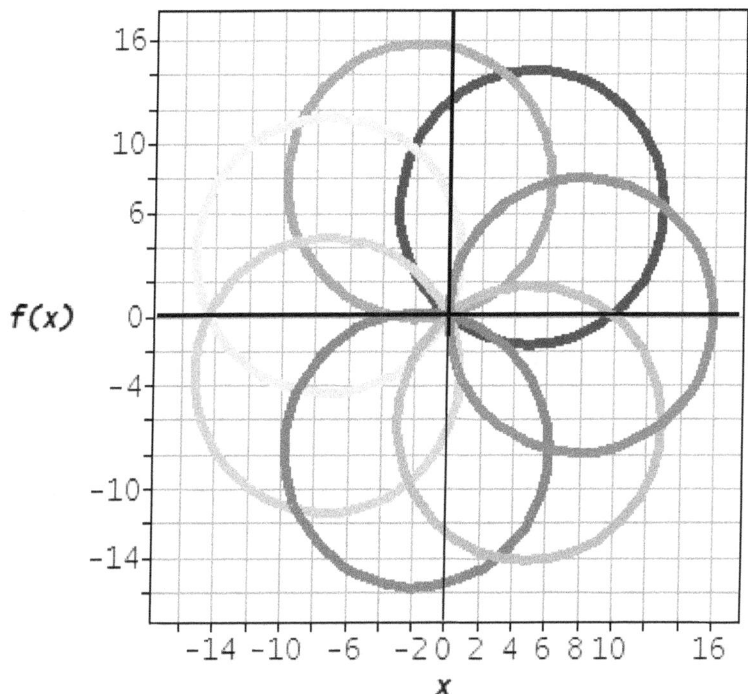

Polygons may be constructed from these circle drawings with an un-ruled straightedge. Polygon constructions could be drawn by compass and straight-edge for up to a septa dodecagon, or seventeen sided one.

www.ingramcontent.com/pod-product-compliance
Lightning Source LLC
Chambersburg PA
CBHW070959180526
45168CB00003B/1206

* 9 7 8 1 5 1 2 3 4 5 6 1 2 *